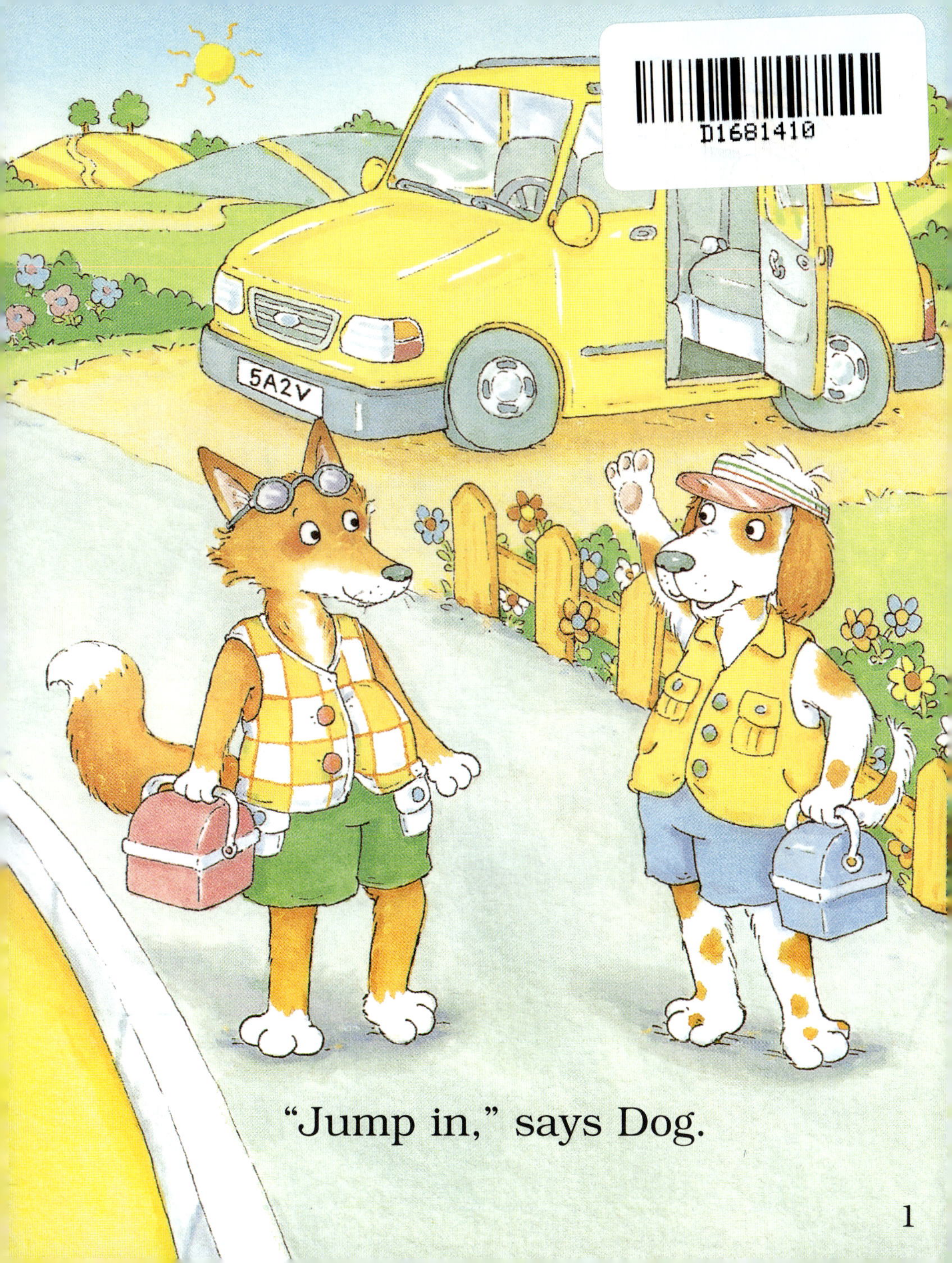

D1681410

"Jump in," says Dog.

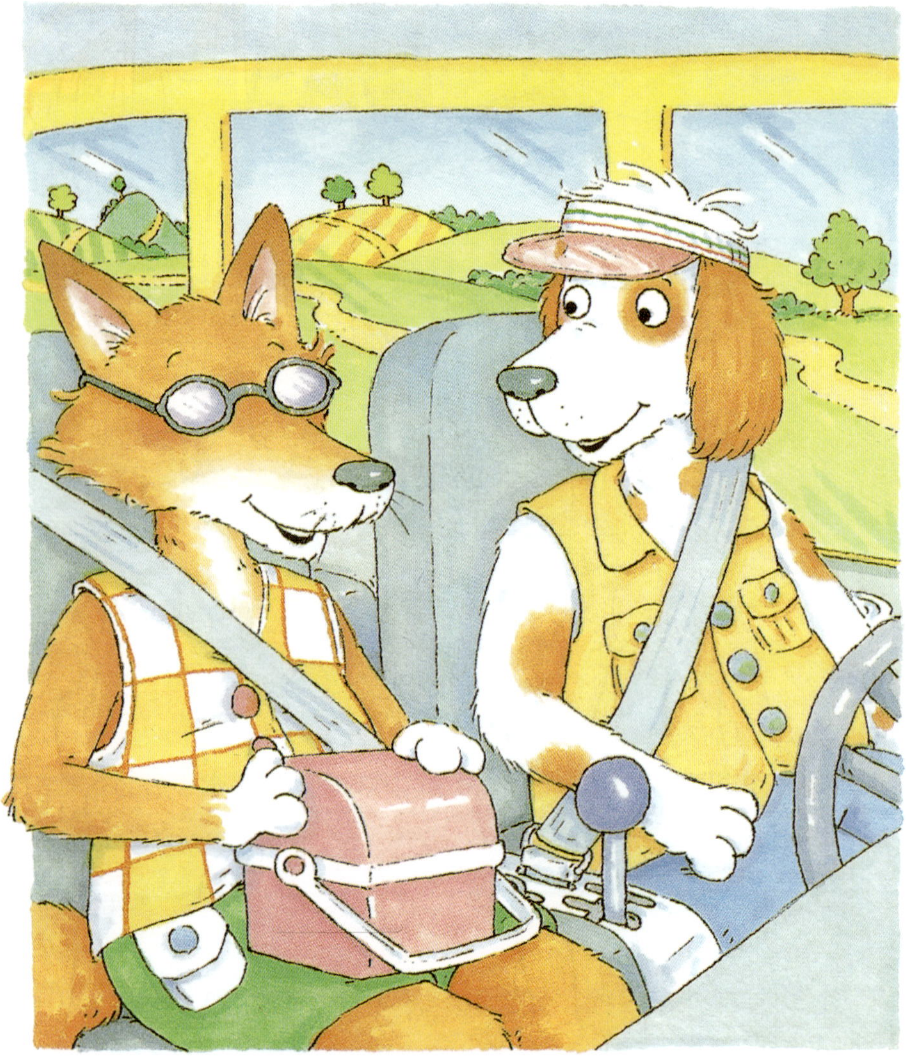

Two go off in the van.

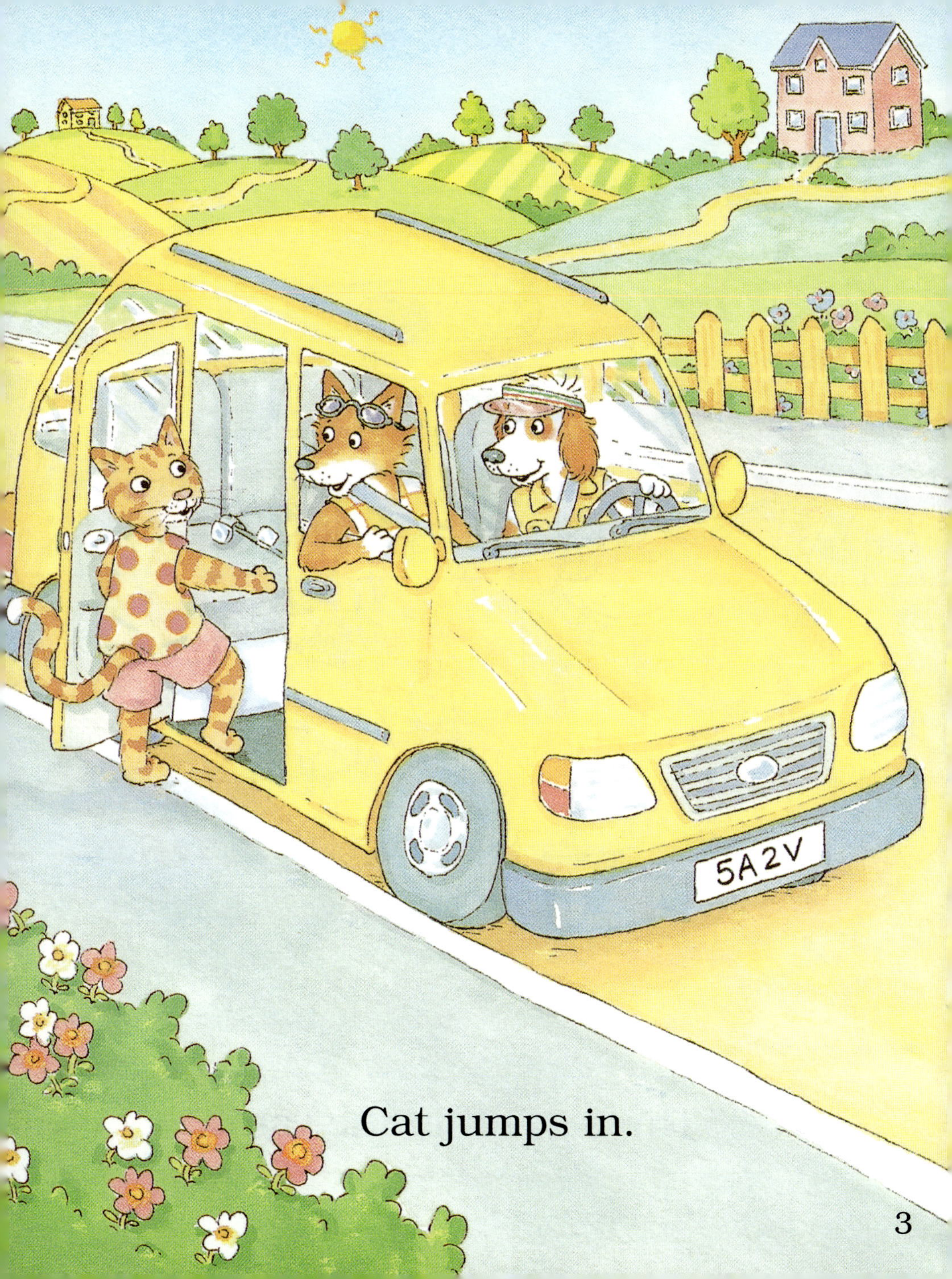

Cat jumps in.

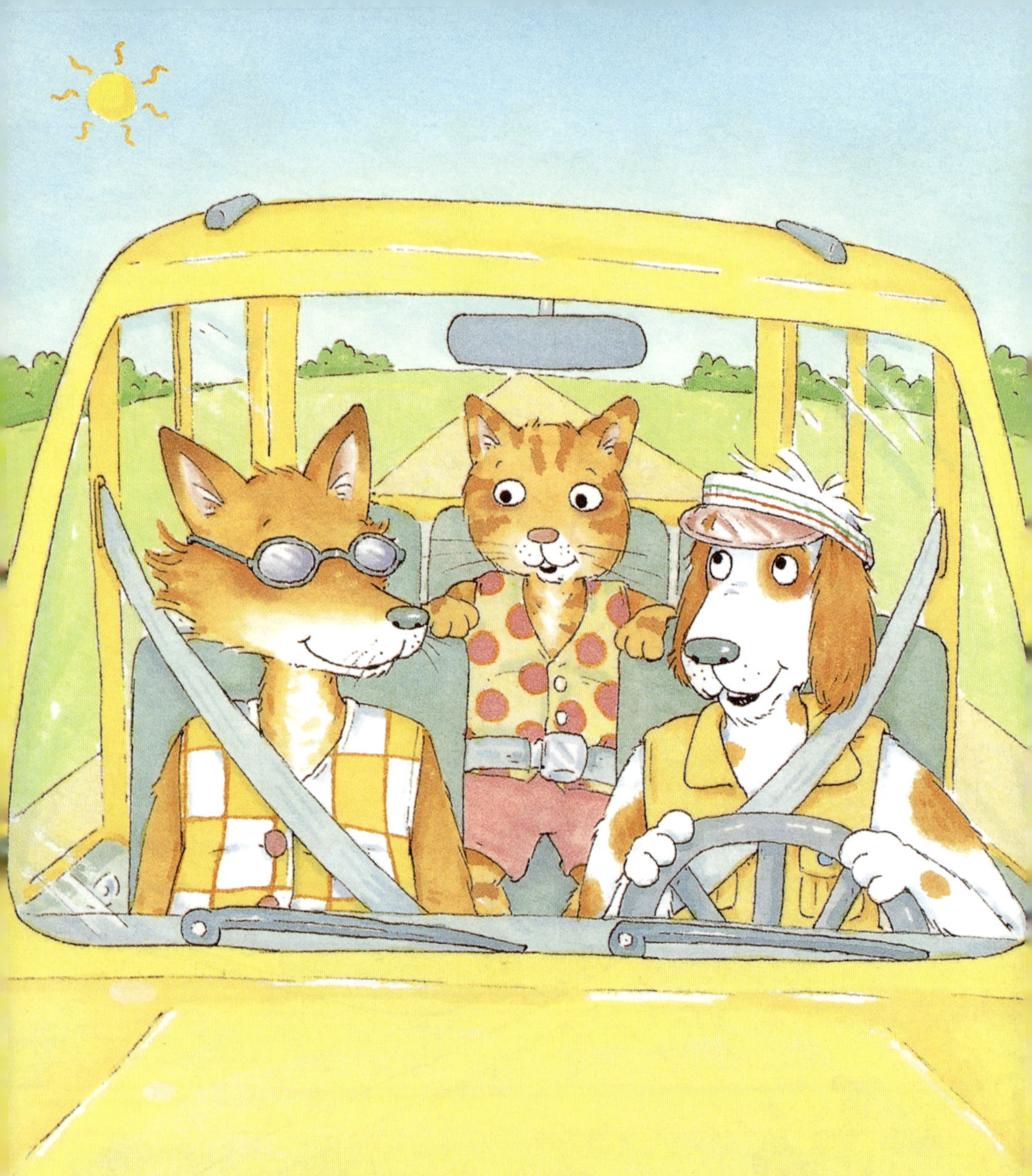

Three go off in the van.

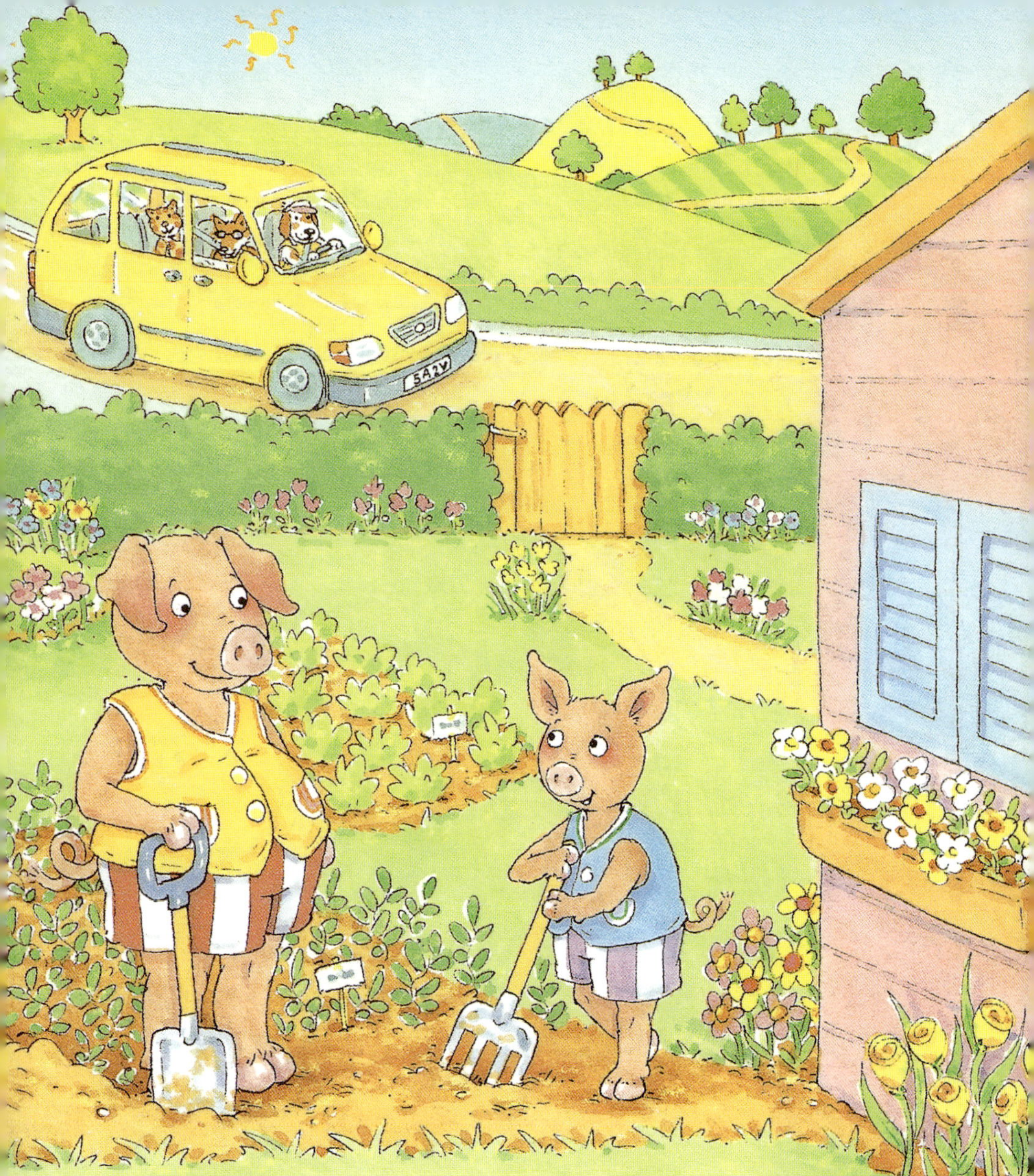

"Hog and Pig live here," says Fox.

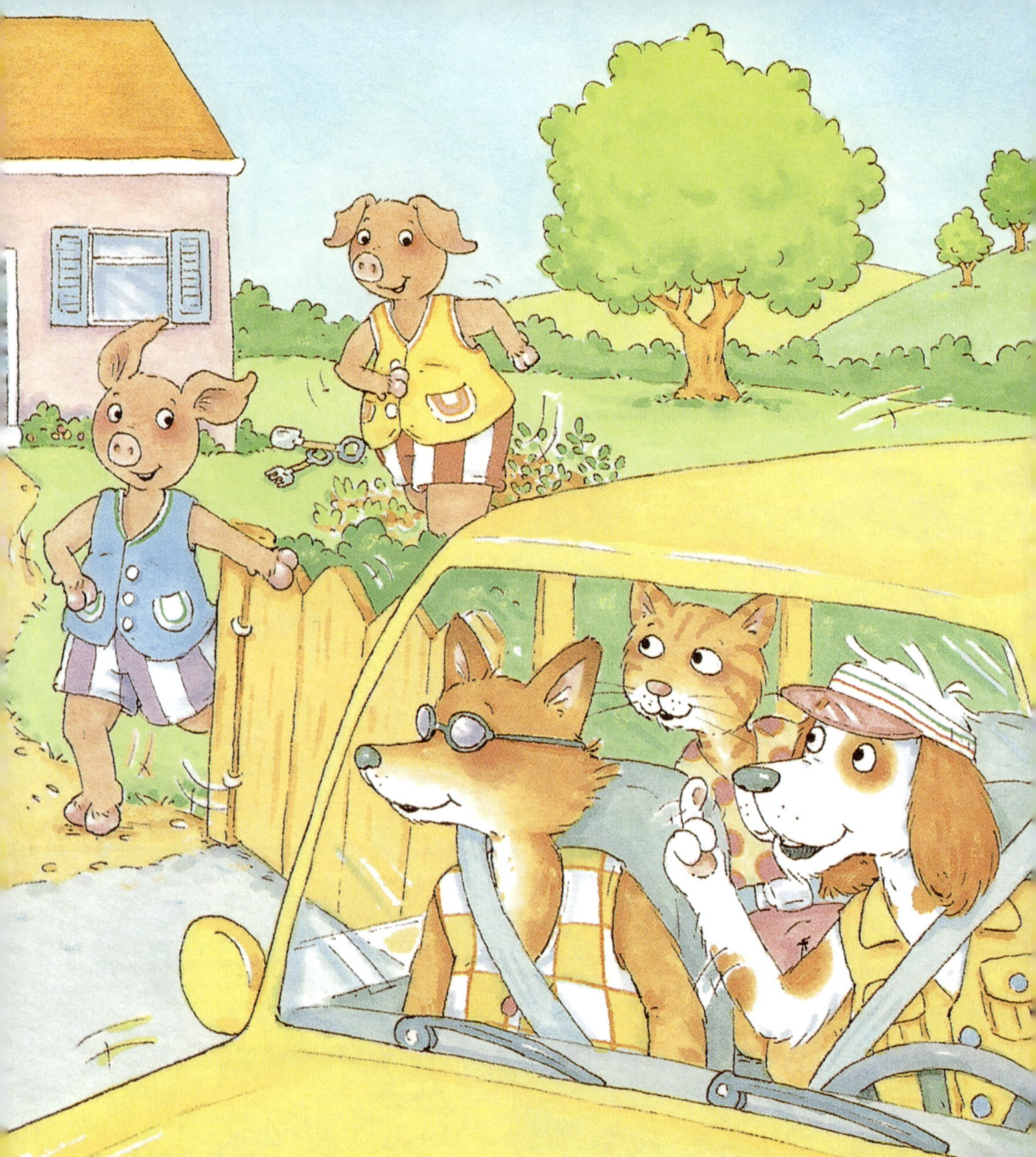

“Jump in!” says Dog.

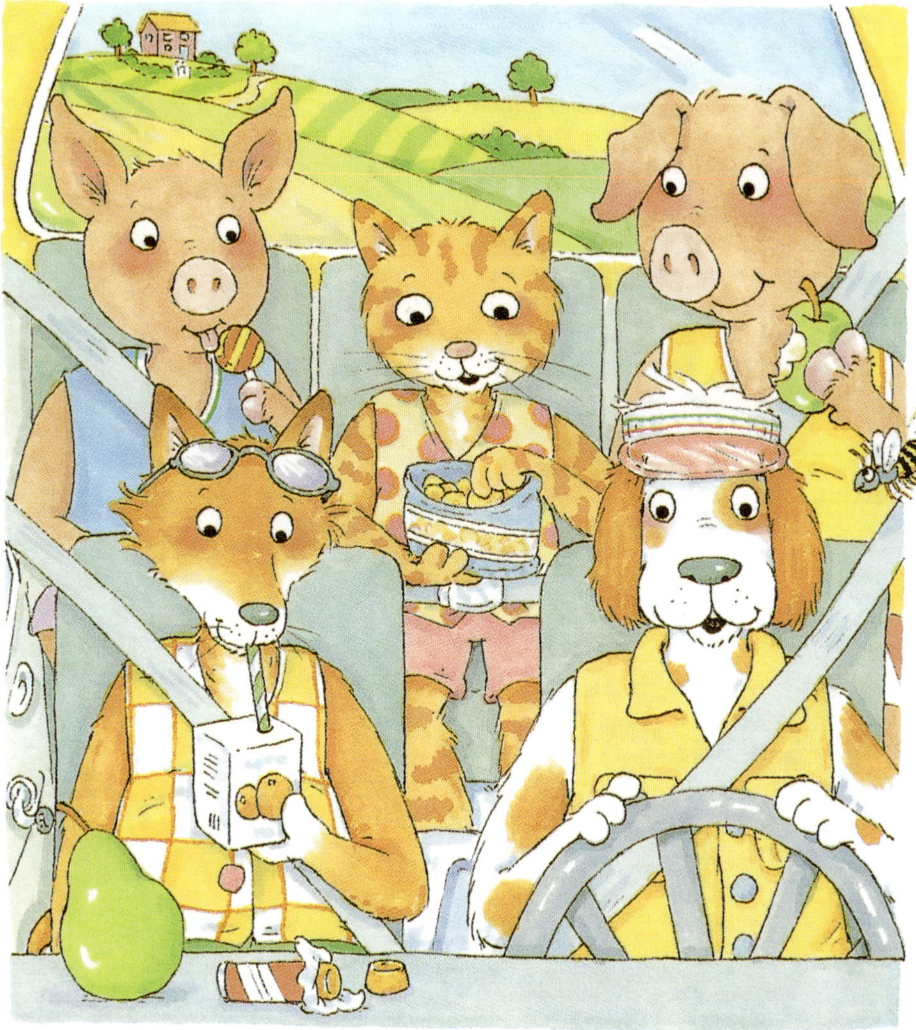

Five go off in the van!

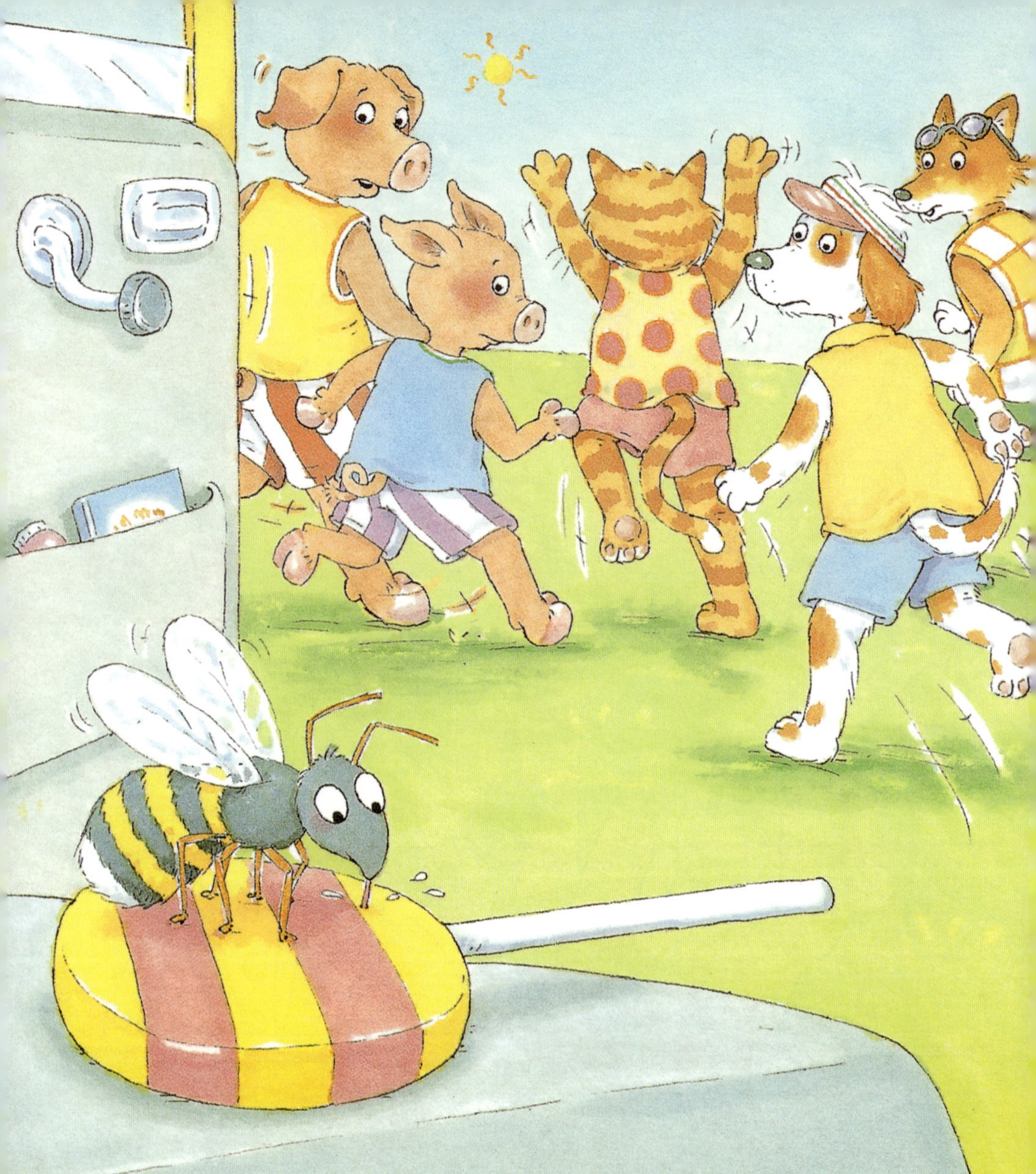

Five jump off the van!